Science, Technology and Development:

The Politics of Modernization

by Thomas W. Wilson, Jr.

CONTENTS

	Foreword	3
1	A Global Fever	5
2	'Progress': The Transformation of an Idea	12
3	Rich and Poor: The North-South Dimension	30
4	Development Strategy Transformed	39
5	Planetary Politics in Its Infancy	48
6	Science, Technology and Politics: Managing Modernization	56
	Talking It Over	61

HEADLINE Series 245, August 1979

$2.00

Cover by Design Works

The Author

THOMAS W. WILSON, JR. has worked on, and written about, international problems as journalist, government official, study director and international civil servant. In the early 1970s he was director of the Program in Environment and Quality of Life of the Aspen Institute for Humanistic Studies. In 1976-77 he served as principal officer in the Office of the Secretary-General of the UN. He is now a consultant to the Presidential Commission on World Hunger. He is the author of three books and has published monographs recently on the environment, energy, food, population, growth policy, and human rights.

The Foreign Policy Association

The Foreign Policy Association is a private, nonprofit, nonpartisan educational organization. Its purpose is to stimulate wider interest and more effective participation in, and greater understanding of, world affairs among American citizens. Among its activities is the continuous publication, dating from 1935, of the HEADLINE Series pamphlets. The authors of these pamphlets are responsible for factual accuracy and for the views expressed. FPA itself takes no position on issues of United States foreign policy.

The HEADLINE Series (ISSN 0017-8780) is published February, April, August, October and December by the Foreign Policy Association, Inc., 345 East 46th St., New York, N.Y. 10017. Chairman, Carter L. Burgess; Editor, Wallace Irwin, Jr.; Associate Editor, Gwen Crowe. Subscription rates, $7.00 for 5 issues; $13.00 for 10 issues; $18.00 for 15 issues. Single copy price $1.40. Discount 25% on 10 to 99 copies; 30% on 100 to 499; 35% on 500 to 999; 40% on 1,000 or more. Payment must accompany order for $5 or less. Second-class postage paid at New York, N.Y. Copyright 1979 by Foreign Policy Association, Inc. Composed and printed at Science Press, Ephrata, Penn.

Library of Congress Catalog No. 79-53795
ISBN 0-87124-055-6

Foreword

Science (the study of the physical universe) and technology (our skilled uses of whatever we learn about that universe that serves our purposes) began to join together only a very few centuries ago in what was to become a uniquely fruitful marriage. From it have sprung, especially in our century and above all in the past generation, transformations of almost every aspect of life in ever-widening regions of the world with the spread of urban-industrial-technological civilization.

For national governments—and above all for the most successful of all high-technology nations, the United States—the application of science and technology today presents far-reaching questions of public policy both at home and in the international realm. One of the greatest of such questions concerns how the benefits of science and technology should be shared—and their undesired effects avoided—through cooperation between the highly industrialized "haves" and the less-developed "have-nots" which now exert substantial influences in the community of nations. Public discussion of this "North-South" aspect of the technology question has been greatly stimulated by the decision of the United Nations to convene a Conference on Science and Technology for Development (UNCSTD)—the first world conference of governments ever held on this subject. The conference, scheduled to meet in Vienna, Austria, in August 1979, was in final stages of preparation as this issue of the HEADLINE Series was being written.

As Mr. Wilson's title indicates, his subject is not so much science and technology in themselves—the history and description of which form an enormous literature—as it is their effects on the development of nations, and of the international system, in the modern age. His concern is especially to explore how deliberate action by governments and by scientific and other bodies may better serve to control or influence those effects for desired purposes. He brings within the range of his discussion problems of science-and-technology policy not only in the North-South context itself but also within the United States—the character of whose future internal development will certainly influence the international scene.

In the sense that Mr. Wilson's essay concerns governmental policies and the distribution of benefits among competing groups, its subject is clearly political. But the issues that technology presents for settlement in the political arena are also relevant to human values and to the quality of life in future generations. It is appropriate, therefore, that this should be among those issues of the HEADLINE Series whose publication has been assisted by the National Endowment for the Humanities, whose support is here acknowledged with appreciation.

—The Editors

1

A Global Fever

Somewhere near the heart of the contemporary world predicament is the fallout of science and technology upon human society. In a mere three to four decades since World War II—quite suddenly, on the scale of historical time—virtually the whole planet has been caught up in a process of technological innovation and social change that goes by the name of modernization.

The modernization process began as the Industrial Revolution in a single region—Europe and North America—about two centuries ago. It spread to Japan in the late 19th century and proceeded to invade Latin America, the Middle East, Asia and Africa with increasing momentum in the 20th. History may record that it became a truly global phenomenon only in 1978, when it was at last decisively embraced by the rulers of China, the world's most populous nation. The picture that now begins to emerge is one in which not a single state in the world system, and scarcely a single cultural group or tribe even in the remotest islands or valleys of the earth, remains completely untouched by modern technology and the fever of modernization.

The wellsprings of modernization are not one but several. They range from the probing curiosity of the scientist and the ingenuity of the engineer and inventor to the entrepreneur's quest for profit and the consumer's appetite for a more comfortable life. These impulses are mutually reinforcing, and they build upon ever-expanding structures of knowledge, technique and organization.

But there is one more key actor in the drama. Modernization has coincided in history with the heyday of the nation-state, the chief locus of political power in the modern age. So potent a process as modernization inevitably has become, in several senses, quintessentially political. It may serve the interests of national governments in various ways. It can reinforce their power and authority, military or otherwise. It also can shape new political issues involving equity and conflicts of value—issues around which much of the contest for political office is waged in most societies.

Three Recent Cases

The far-reaching political implications of the modernization process can be illustrated in strongly contrasting ways by a quick glance at certain recent events in three countries: China, Iran and the United States.

In *China,* the first generation of revolutionary leaders who took power in 1949 conducted a long, hard struggle against the infectious ideas of modernization. Especially after the ideological split with the Soviet Union around 1960, the People's Republic lived in self-imposed isolation. The overriding national economic goal was maximum self-reliance. Technical innovation was passed over in favor of traditional methods of production. Machinery, except for modest programs in heavy industry and nuclear weapons, was generally shunned in favor of mass manpower. Improved living standards would depend mainly upon a more evenhanded distribution of goods and services.

These policies, however, were not unanimously supported within the Chinese ruling establishment. In fact, the issue of modernization was a major source of factional struggle. The Cultural Revolution

(1965-68) was the high tide of the antimodernizers, and its excesses were major counts in the allegations against the Gang of Four after the death of Mao Zedong.

The successors of Chairman Mao lost little time in turning things upside down. In 1978 there began an all-out campaign to modernize agriculture, industry, the armed services and the scientific and technological resources of China by the year 2000. It was an ideological turnabout of 180 degrees, and the *People's Daily,* official newspaper of the Chinese Communist party, made no bones about it. In a major editorial about the "big push in the modernization drive," the paper stated: "Modernization will be the central task of the whole party from now on. All other work, including the party's political work, will focus on and serve this central task." What's more, the editorial warned: "There must be no 'political movement' or 'class struggle' which deviates from this central task."

It cannot be known by an outsider how much of its public commitment to all-out, forced-draft modernization the present Chinese leadership actually expects to carry out. There are legitimate doubts as to whether enough capital can be found to go so far so fast as the first ambitious announcements suggested. In fact, these already have been revised and there have been reports of canceled contracts for development projects. But there can be no doubt of the direction in which China is moving. Mechanization will increasingly replace labor-intensive methods of production. Economic isolationism will begin to yield to more active interdependence with the world outside. China, with nearly a quarter of mankind, is launched upon a course of profound technological modernization and social change that is likely to preoccupy the time and attention of its leaders for decades to come.

Even as China was cranking up its "big push" toward modernity, rioters in the city streets of *Iran* were bringing down a regime which had appeared to hold all the cards of conventional power. Plainly the Iranian revolution has multiple roots and mixed motivations which cannot yet be sorted out with confidence. But just as plainly, the events in Iran were, among other things, a protest against modernization—at least against the kind of development and the pace of social change that were taking place in that society. To some significant segment of the Iranian population, modernization meant "Westernization"—and an affront to valued tradition.

Some Iranians warned of trouble years before the eruption. At a symposium in Iran in 1975, Hormoz Farhad, vice-chancellor of the University of Farabi in Tehran, expressed deep misgivings about the pervasive contemporary search for technological advancement. Modernization in Iran, he said, "is coming about swiftly and mostly as a result of blind imitation of what is happening in the West." Rising incomes, he said, had led to the "emergence and constant growth of a middle class"—to which he added: "This middle class is displaying a gradual moral breakdown that is quite alarming. Respect for one's faith, for one's fellowman, for decency, honesty, and the law has been diminishing. The resultant spiritual bankruptcy is perhaps the most serious threat to the very fabric of the Iranian society."

Three years later—in the midst of the shah's frenetic race to transform an ancient society into a modern industrial nation and military power within a single lifetime—the fabric of Iranian society was ripped into shreds. Nobody could explain exactly how it came about or just where it would all end. But in *The New Yorker* magazine of February 19, 1979, William Pfaff wrote that "Iran has proved to be a model of how a developing country goes into shock as a result of economic development, social change and the resulting attack upon its traditional culture."

Meanwhile, in the *United States* the modernization process was running into a different kind of problem. Technology was coming to be seen not only as a means of solving problems but as part of the modern predicament. In some areas, technological innovation seemed increasingly random, undirected and frequently frivolous. In some respects, industrialization appeared to have run amok. The very idea of progress began to come under question.

These troubling notions gained frequency through the 1970s. Then they rose to high drama early in 1979 in what the Nuclear

Following the nuclear accident at Three Mile Island, nearby town is practically deserted as officials check area for radioactivity.

Regulatory Commission calls "unscheduled events in the fuel cycle." In the early morning of March 28, in the control room of Reactor 2 at the Three Mile Island nuclear generating facility near Harrisburg, Pennsylvania, bells began to ring and lights began to flash on the 40-foot-long panel of supermodern instrumentation. While engineers were trying to figure out what had gone wrong, the computer monitoring temperatures in the dome of the reactor stopped recording numbers and began printing question marks.

It may be years before the impact—psychological as well as technical—of this accident upon nuclear energy policy can be assessed clearly. Early assessments suggested that its implications for public opinion might well prove profound, and would not stop at national frontiers. As a placard carried a few days later in an antinuclear demonstration in Germany said, "We all live in Pennsylvania."

Time of Transition

In retrospect it may appear that the main casualty in the events at Three Mile Island—and in the thousands of other environmental disasters, large and small, that have troubled our technological generation—was the excessively simple American idea of ever-advancing technology as the automatic guarantor of everlasting progress. That idea is close to the center of a cluster of assumptions and values that have dominated public and private decision-making through most of the history of the United States. It came unraveled gradually during the 1970s as government was called in again and again to correct technological missteps which it had largely failed to anticipate and which the free market seemed unable to correct on its own. Now the United States and, to some extent, other industrial nations, appear to be in the early stages of a major transition from indiscriminate economic growth to more selective directions of growth.

From these three very different samples of the interaction between technology and politics, we may draw a reminder of how deeply every nation, no matter what its condition, is involved in technological questions. But each of the three examples also teaches a special lesson. In China we see a nation to which modern technology is still largely alien, but which in the end could not resist its advantages. Technology is power—to feed and heal, to build a nation, to defend against enemies. No political leadership, of any ideological persuasion, can remain immune to the promises of modern technology.

In Iran, on the other hand, we can discern a warning that forced-draft modernization can breed a potent resentment from those whose sense of traditional values and feelings of continuity are offended—and from those who find the distribution of benefits to be inequitable. And in the United States we see how powerful technologies, insufficiently directed by public policy toward chosen goals, can reveal unsuspected hazards, arouse public anxiety, and confront policy-makers with new and agonizing choices between economic growth and a healthy environment.

Many of the deepest problems of the role of science and technology in modern society are unresolved. Can progress be redefined in terms more humanly meaningful than annual growth in the gross national product (GNP)? Can rapid technological change be reconciled with man's profound need for continuity? Can the benefits of technology be shared in ways that democratic societies will increasingly accept as equitable?

These are questions that every modern nation must face. What is more, the same questions must be faced at the global level. The operations of multinational business, international scientific and engineering associations, highly skilled individuals moving from nation to nation, the manifold interactions of governments—all these in combination have jumped over national borders and made the entire world a single arena for the uses of science and technology in the universal drive for modernization, the conquest of poverty and the improvement of society. Yet "the world" cannot take action. The ultimate repositories of political power in our fragmented world—about 160 national governments, large and small, at all stages of social and technological evolution—have only begun to grope their way toward an international order coherent enough to settle the countless issues that arise in applying the power of science and technology to constructive ends.

A new stage in that groping process was reached with the launching of the United Nations Conference on Science and Technology for Development (UNCSTD), Vienna, August 1979. In a later chapter we shall turn to some of the issues with which that conference must deal, issues which in a deeper sense will be with us for decades to come.

2

'Progress': The Transformation
of an Idea

The early Egyptians, Chinese, Greeks, Hindus, Persians, Arabs and other ancient peoples thought hard about the world around them, and came up with discoveries and ideas that led to the accumulation of a vast world stock of knowledge. But none of these early civilizations achieved the basic principles of the scientific method—rigorous observation of cause and effect and demonstration of general laws through controlled, repeatable experiments. These came only with what we call the scientific revolution, which began in the 16th and 17th centuries in Europe. The Industrial Revolution, the age of steam and steel, dates back only about 200 years. And the so-called knowledge explosion—the systematic, large-scale use of research and development to devise new products and processes—has happened within the past three decades. So the "march of science" has been swift and accelerating, and the proliferation of technology at the rates we know today is a very recent phenomenon. As for the *politics* of modernization, it is only now being explored for the first time.

It is sobering to recall that scientific discovery and technological innovation, for all their enormous influence, have been opposed or deplored at every step of the way. Four main themes can be identified in this chorus of opposition: science vs. religion; technology vs. nature; machine vs. man; and modernization vs. tradition.

Science vs. Religion

The beginning of systematic scientific investigation confronted church authorities about four centuries ago with new sources of knowledge which implicitly challenged religious scriptures and their explanations of the universes of man, nature and God. Science offered a new and nontheological way of searching for truth— indeed, a new epoch in the evolution of human thought. So the Christian theologians who fought back knew what they were doing. Their names include not only the half-forgotten popes and grand inquisitors who condemned Galileo but more famous names, both Catholic and Protestant: Martin Luther, John Calvin, Cardinal Richelieu, Sir Thomas More, John Wesley and Cotton Mather, among many.

Notable, too, are the recurrent attempts of some fundamentalists or pietistic religious movements to limit or forbid the advances of technology. "What can you conceive more silly and extravagant than to suppose a man racking his brains, and studying night and day how to fly?" wrote the evangelical author William Law in 1828; and in another century a widespread saying went: "If God had wanted man to fly he would have given us wings." In every major religious tradition there are ascetic or pietistic sects whose members deny themselves what they consider the sinful luxuries that flow from technology. And, as was dramatically demonstrated in Iran in 1978, when modern technology is force-fed to an ancient culture which perceives it as alien, the reaction against it can take a religious form—with explosive effect.

Fundamentalist and pietistic strains of religious faith are not dead, then—in Christendom or in Islam. They remain a widespread, if often self-defeating, manifestation of man's need for

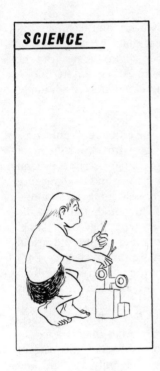

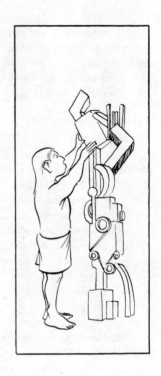

continuity and certitude in an age of bewildering change. They are, however, only one of the responses that churches and religious thinkers have made to the ascendancy of science in our age. Even before 1700, the great astronomer Sir Isaac Newton saw God's hand in the cosmic harmonies his scientific genius had discovered. Today, many modern-minded theologians and moralists strive to reconcile their traditional creeds with the revelations of science, and to make technology serve the moral ends their religion teaches. "Gradually, slowly, steadily," wrote the philosopher-mathematician Alfred North Whitehead in *Science and the Modern World* (1925), "the [religious] vision recurs in history under nobler form and with clearer expression. . . . The fact of the religious vision, and

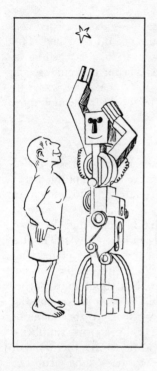

Vadillo in *Siempre,* Mexico

its history of persistent expansion, is our one ground for optimism."

Whether the encounter between science and religion ultimately will prove destructive or creative, only time can tell. However, a two-volume history by Andrew Dickson White, published toward the end of the last century, records in great detail the story of the effort of orthodox Christian theologians to destroy not only scientific explanations of the creation of the world and the evolution of species, but new theories and findings in geography, geology, archaeology, ethnology, meteorology, chemistry, physics, medicine, hygiene, psychology, philology, mythology and political economy.

The author—while insisting that the struggle of science was not with *religion* but with dogmatic *theology*—also insisted on proclaiming total victory for science and the scientific method. The judgment was premature. Decades later, in 1925, the teaching of Darwin's theory of evolution was adjudged a crime in Tennessee in the famous "monkey trial." As late as 1977 the issue was still reverberating in a fight over public school textbooks in the mountain retreats of West Virginia.

Technology vs. Nature

From the beginning of the Industrial Revolution, a procession of poets, painters, novelists and social critics have warned of the threat posed by machines to the beauty of the countryside, to the values associated with rural life, and to the integrity of natural systems and resources. In England, the poets William Blake and William Wordsworth, the novelist Charles Dickens and the essayist Thomas Carlyle, among others, inveighed against the squalor of teeming cities and, in Blake's phrase, the "dark, satanic mills" of the industries spawned by new technologies. Nathaniel Hawthorne, Ralph Waldo Emerson and Henry David Thoreau are among the American writers of the 19th century who took up the theme of tension and conflict between the world of rural peace and simplicity and the world of urban power and sophistication—in a literary vein which can be traced back to Virgil's celebrations of the virtues of pastoral life under the Roman Emperor Augustus.

In the political realm, the impact of industrial technology upon nature provided a major motivation for the American conservation movement. Even before the turn of the present century the movement had enough public support to persuade Congress to make Yosemite the first national park in a system which has undergone its greatest period of expansion within the past few years.

Machine vs. Man

The deep concern of the romantic school of writers and painters about the intrusion of technology into nature and rural life is paralleled by another literary theme: a sometimes morbid fascination with the relationship between human beings and the machines they invent. Mary Shelley's novel *Frankenstein,* with its horrifying yet pitiable monster escaping the control of its scientist-creator, was published in 1818, just as industry stimulated by the Napoleonic wars was beginning to transform England. A century later, Charlie Chaplin's film *Modern Times* was one of many works raising the same fearful question: whether man is, in practice, master of the machine or vice versa.

On a more mundane level, the machine has appeared as man's enemy whenever it has abolished jobs on which workers depend for their living—or when it enabled the few to prosper by the ill-paid toil of the many. Indeed, it was revulsion against the abysmal poverty and wretched conditions of factory workers in Europe that inspired Karl Marx and Friedrich Engels in 1848 to write the *Communist Manifesto* and later led to social reform, acceptance of trade unions, and eventually to the welfare state. Yet this source of tension still persists today. Labor unions bargain and go on strike over the same issue that led the Luddites to smash up textile machinery in 19th-century England: machines replacing humans in the workplace. And there is no chance that we have heard the last of that story.

Modernization vs. Tradition

Those who resisted the surging growth of science and technology were up against a politically powerful set of ideas. Science not only offered a source of knowledge apart from divine revelation. Translated increasingly into useful technology, it also undercut the age-old fatalistic assumption of traditional societies that the masses of mankind were destined to live out their lives in material misery. Science and technology offered not only an alternative mode of thought but a potentially alternative mode of life based on the revolutionary idea of "progress."

The basic notion was that the material components of the human condition can be improved right here on earth—that surcease from suffering need not await an afterlife. This idea drew strength from a

Sweatshops were common in England and the United States during the Industrial Revolution. Picture of this necktie factory in a New York tenement was made by the American reformer Jacob Riis.

companion idea, which became increasingly demonstrable as the Industrial Revolution gathered speed in the early 18th century: that technology applied to production results in economic expansion, a steady growth in the volume of product wrested from natural resources for the use of man.

From this new awareness, in turn, came a question which was to shape much of the politics of the modern world from then on. Who is to benefit by this growth: the few or the many? Contention over this great political question was one of the main forces leading to the expansion of the franchise and the spread of democracy on both sides of the Atlantic in the 19th and 20th centuries. And from this contention there also flowed a further powerful idea: that by the deliberate policies of government, economic growth can be consciously *directed* to improve mass standards of living. Thus, from the beginnings of the Industrial Revolution to the present day, the political leadership of country after country began to focus on the task of national development—on the modernization of national societies through constant growth, heavily dependent on technology and wholly dependent on a vision of the possibility of progress.

Science . . . technology . . . progress . . . growth . . . development . . . modernization: this pattern of interlinked ideas became a central part of the operative value structure in Western, modernizing societies. It was incorporated in national policies. It influenced the shape and missions of important institutions, government agencies, universities, foundations, business, labor and professional associations. It tipped an endless number of public and private decisions in one direction or another. And, lest the point be overlooked, these ideas made possible the historical phenomenon known as the consumer society.

In countries undergoing industrialization, the modernization process became the driving force and the organizing principle of political action. And in that framework, economics and engineering would become dominant influences in political decision-making.

In the Western world, and especially perhaps in the United States, a prevailing faith developed in the "endless frontier" of modern science; in the scientific method as the best path to dependable truth; in the scientific mind as the ultimate agent for the solution of almost any problem that can be formulated; and in the notion that science and technology promise limitless progress. Technological optimism became a prevailing frame of mind.

Quite recently, some historians and others have begun to explore the theory that modernization may be at the very heart of the historical process itself. Surprisingly, this theory seems to have been pushed farthest among some of the more unorthodox Marxist intellectuals of Eastern Europe. There the view has been propounded that history is essentially the record of societies adopting whatever technologies are available to them, and that the overwhelming historical force in the contemporary era is not the class struggle, as classical Marxism asserts, but the scientific and technological revolution.

Technological determinism is as unsatisfactory a theory as any other kind of historical determinism. But one does not have to accept the notion that science and technology are the exclusive shapers of historical process to see technology as a prime generator of massive and pervasive social change; nor to see that the impact of science and technology would fall upon whole societies.

Doubts and Questions

In our time it was the anthropologists, studying societies largely untouched by the idea of progress, who were among the first to notice the society-wide dimensions of modernization—and to raise some disturbing questions.

Writing a quarter of a century ago, the late Margaret Mead formulated dilemmas which only now are being heard in the political corridors of the world community. "The words technical change," she wrote, "have come to symbolize for people all over the world a hope that is new to mankind—the hope that the peoples of the world need be hungry no longer." Then she added at once: ". . . granted that we know the technical answers . . . what will be the cost in terms of the human spirit? How much destruction of old values, disintegration of personality, alienation of parents from children, of husbands from wives, of students from teachers, of neighbor from neighbor, of the spirit of man from the faith and style of his traditional culture must there be? How slow must we go? How fast can we go?"

There are no good answers to these questions today. There are no accepted general theories of the modernization process, no models for studying the behavior of societies under the stress of change. But the questions are more pressing today than when they were first asked. As Princeton historian C.E. Black sees it:

"The process of change in the modern era is of the same order of magnitude as that from prehuman to human life and from primitive to civilized societies; it is the most dynamic of the great revolutionary transformations in the conduct of human affairs." The critical thing for political leaders now, Black believes, is to distinguish "what must unavoidably be changed, and what must at all costs be preserved."

This is a central issue for political leaders almost everywhere—and not least of all, in the United States.

For it can now be seen that for a quarter of a century after World War II, the United States and other industrialized countries were so intent on the unprecedented benefits of new technology that its dangerous and destructive side effects were scarcely noticed. Medicine took dramatic steps toward the conquest of disease. The computer began relieving mankind of clerical and computational work just as the industrial machine had replaced physical labor; more dramatically still, it ushered in automatic, split-second control over immensely complex processes, from auditing the nation's tax returns to landing men on the moon. Communications technologies offered dazzling prospects at home and abroad.

But then suddenly, it seemed, physical evidence of the unintended side effects of rampaging technology was everywhere. Streams, rivers and lakes were becoming polluted with industrial and agricultural wastes; estuaries and wetlands were becoming contaminated or disappearing to make way for "development"; prime agricultural land was being buried under superhighways; even the high seas, which cover 70 percent of the planet, were showing the strains of being the "ultimate sink" for the waste products of civilization. Smog over urban areas became a familiar sight from the air—while carbon dioxide from accelerated burning of fossil fuels and particles of matter rising from factory smokestacks fouled the quality of air in cities where most of the people live. The waste products of technological civilization that were not dumped into waterways or the atmosphere were dumped on the land—sometimes in random disarray, like the ubiquitous automobile graveyards and the thousands of places in the countryside where millions of drums containing toxic chemical wastes were abandoned to work their corrosive ways back into the environment.

There were unwanted social by-products too. By the 1960s some members of the "advanced" societies were tormented by the feeling that things were "out of control"—that their institutions had lost the capacity for governance. Crime rates, addiction rates, divorce rates and alienation were cited as evidence. Whatever the causes of the unmistakable social malaise, technology—or at any rate the abuse of technology—received the lion's share of the blame from some scholarly critics.

After a lifetime of studying the interactions of technology and civilization, American social critic Lewis Mumford reached a glum conclusion: "Nothing less than a profound reorientation of our

Victim of air pollution in Japan receives oxygen.

United Nations/WHO/Takahara

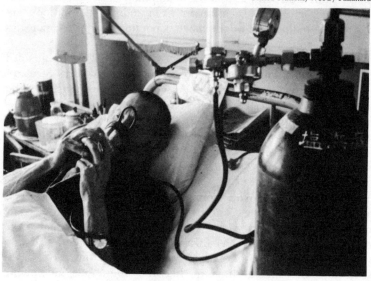

vaunted technological 'way of life' will save this planet from becoming a lifeless desert."

The renowned microbiologist and essayist René Dubos put aside his customary optimism to discount the prospect of technological solutions to contemporary social problems. "Technological fixes," he wrote, "usually turn out to be a jumble of procedures that have unpredictable consequences and are often in conflict with natural forces."

From France, sociologist Jacques Ellul asserted: "Technique has become autonomous; it has fashioned an omnivorous world which obeys its own laws and has renounced all tradition."

Daniel J. Boorstin, historian and Librarian of Congress, concluded in a bicentennial essay for *Time* Magazine, that technological change is binding the world together "with crushing inevitability" into a community of shared experience—a "Republic of Technology." But the costs would be high. For, as Boorstin warns, the Republic of Technology "is a world of obsolescence" in which "technology and advertising create progress by developing the need for the unnecessary." Moreover, experience in the Republic of Technology "uproots us and separates us from our own special time and place . . . we are protected from the climate, the soil, the sand, the snow, the water. Our roots, such as they are, grow in an antiseptic hydroponic solution." And in what might be taken as a warning to the rest of the world on the occasion of the U.S. bicentennial, Boorstin suggests that the Republic of Technology, like the United States, would be a "world whose rhetoric is advertising, whose standard of living has become its morality . . ."

Society Reasserts Control

Yet even as the antitechnology and antiscience literature was accumulating during the 1960s, so was the evidence that society was moving to assume control over the juggernaut of technological change. Operative ideas and values that had dominated decision-making in the industrial world for two centuries began to be called into question not just by social critics but by policy-makers. The idea of progress began to be redefined to embrace the quality of life as well as the quantities of goods produced and consumed.

It is possible to assign an approximate date to this change: 1970, just 25 years after the end of World War II and the release of the knowledge explosion. In that year, the National Environmental Policy Act (NEPA) became part of the law of the land of the United States. The full implications of this landmark legislation may not be felt for another decade or so, and it almost surely will be at the center of coming political storms.

Enactment of the NEPA was an explicit response to a phenomenon which even today has not yet been grasped fully in or out of government. The new law noted "the profound impact of man's activity," including population growth, urbanization and other aspects of the modernization process, "on the interrelations of all components of the natural environment." It laid down a new set of basic principles—among them that the Federal government has a "continuing responsibility" to act in furtherance of "the responsibilities of each generation as trustee of the environment for succeeding generations"; the need to "preserve historic, cultural and natural aspects of our national heritage"; and the objective of "maximum attainable recycling of depletable resources." These new social goals involved a clear departure from the notion of indiscriminate economic growth for the indefinite future. To give effect to these goals, the law imposed some radically new requirements.

Possibly most far-reaching among these requirements is one imposed on all agencies of the Federal government whenever they propose an activity that could have a significant impact on the natural environment. To guard against harmful environmental consequences, the act requires that each program seeking approval be accompanied by an Environmental Impact Statement anticipating all harmful effects of the proposal not only in economic terms but on amenities and aesthetic values as well. Nobody yet knows exactly how to perfect such an analysis. But adoption of this legislative requirement was a first step toward deliberate, advance analysis of alternative ways of meeting economic ends in light of their economic, social and environmental consequences.

Congress soon took a further step in that direction by establishing an Office of Technology Assessment to explore the implications of proposed or prospective technological innovations. The art of technology assessment remains in a primitive state, but the initiation of such an approach asserts an intent by the government of the United States to determine what social ends technology should serve.

The NEPA thus contributed to a reordering of the goals and values of American society in transition. But its reach was more than national. It specifically recognized "the worldwide and long-range character of environmental problems" and called on the United States to join with other nations "in anticipating and preventing a decline in the quality of mankind's world environment."

The NEPA did not stand alone in that year 1970. It was rather the legislative expression of a new national awareness. On April 22 of that same year, Earth Day was celebrated on the campuses of the nation. From that moment the environmental movement became an enduring element of political reality in the United States and, in fact, in all industrial democracies. In the years that followed it would take many forms as it slowly came to embrace not only natural environments but urban neighborhoods and the work-place—the artificial as well as the natural parts of the human environment.

Many books of the time made their contribution to the turn away from indiscriminate growth and change. *Future Shock* by Alvin Toffler (1970) threw a hard light on the phenomenal acceleration in the rate of change, and on the impact of accelerating change upon society and the individual. This was one of many books by which the notion that "change" and "progress" are synonymous was rudely shaken. A so-called "disease of change" and the rate of change became subjects for study and anxious discussion.

Two years later, in *The Limits to Growth,* a report prepared for the Club of Rome, a team of systems analysts headed by Dennis L. Meadows concluded that assuming no change in the trends then prevailing, "the limits to growth on this planet will be reached sometime within the next 100 years," and that "a rather sudden and uncontrollable decline in both population and industrial capacity" would probably result. The moral, they said, was that humanity must start as soon as possible working for a "state of global equilibrium."

The report stirred up a storm of technical protests, loud objections on ideological grounds and an anguished response from the poor, who felt they were being dealt out of the modernization process. But the report's basic message could not be erased. The old assumptions that nature was there for the taking, that resources are effectively limitless, that energy supplies can be taken for granted—these corollaries of the first, primitive idea of material progress—were largely discredited by the mid-70s.

Starting in 1970, important areas of industrial, transport, mining, energy and agricultural technologies were placed under control in the name of environmental integrity—a new, politically defined social purpose. The NEPA was followed by a series of legislative steps mandating national air and water quality standards and other environmental restraints. It would not be long before certain types of aerosol spray cans would be outlawed because of concern that the propellant gas used in them might damage the ozone layer which screens out the lethal ultraviolet end of the spectrum of sunlight. Health and safety standards increasingly constrained the uses of technology. Industrial processes from steelmaking to meat-packing were hedged by environmental standards whose cost was deemed justified as a matter of social policy. Recycling of waste paper, metals and glass became a new national cause. Protection of forests, wetlands, rivers, seacoasts, endangered species and other key ecological resources was given new weight against the claims of sheer economic growth. In all such matters, action by states, localities and private citizens supplemented—and often ran ahead of—the Federal effort.

About the same time, some lines of technological development were running into dead ends. By the decision of Congress in 1971 not to pursue the French, British and Russians into production of a supersonic airliner, the world's leading nation in aircraft manufacture opted out of a technological competition in which not only national prestige but also vast commercial gains were allegedly at stake. The decision dramatized a new national mood in which even that most American of virtues, speed, was no longer felt to be enough to justify the production and use of a product that could outperform its predecessors, even though its technical feasibility was unquestioned. The extra speed was simply not worth the financial and environmental price. Similar considerations later would turn aside or restrain devices ranging from the nuclear breeder reactor to the video-telephone.

Even aesthetic factors were coming to influence the design and uses of technology—as seen in the strip-mining control act; in national, state and local legislation setting aside parks and recreation areas and preserving wilderness and wildlife; in legal battles over the siting of industrial installations; and in mazes of zoning restrictions at the local level which reflect changing ideas about the virtues of indiscriminate growth.

What was happening in the United States in those years was no less than transformation of the idea of progress. It was a transformation well expressed in the words of a county planning official in the Shenandoah Valley of Virginia where residents in 1979 were divided as to whether they wanted a large brewery to be built, despite the jobs and higher salaries it would bring to the valley. "It's not really the brewery itself that is the issue," he said, "it's the growth that will come with it. Where will we put that? Can we control it? Can the community handle it psychologically?"

These would have been strange sentiments only a few years earlier. But by the end of the 1970s they were typical of communities rejecting or restraining projects once sought eagerly in the name of progress.

In those same years, the scientists themselves were beginning to wonder out loud whether the frontiers of science are indeed limitless. In its Spring 1978 issue, *Daedalus,* the Journal of the American Academy of Arts and Sciences, was devoted to the topic "Limits of Scientific Inquiry." The preface asserts that to have explored the idea that there may be limits to scientific inquiry ten years previously "would have seemed inappropriate, even incongruous to a society overwhelmingly preoccupied with the problems created by the orbiting object called Sputnik." Yet, it continued, "questions are being asked today, by scientists and others, that would have been almost inconceivable 20 years ago."

In 1979, there appeared in New York City a new quarterly, *Technology in Society.* In the first issue, an article by the Harvard physicist Harvey Brooks expressed the sober mood of the scientific community with the conclusion "that science is more hope than catastrophe though, admittedly, it is nip and tuck." But although the mood is no longer one of facile optimism, neither is it despair. In an introduction to the same issue, the editors wrote that "of all the factors involved in the day-to-day choices confronting the individual and society in our complex environment, technology itself is uniquely amenable to deliberate human control."

In short, some fundamental ideas were being revised in the United States in the 1970s. The idea of progress was being redefined to embrace something more than quantitative growth of goods and services. There was an increased awareness of limits to the capacity of science to resolve social problems. And there was growing insistence that conscious guidance should replace indiscriminate proliferation of technology.

In retrospect, we can see that the modernization process for the highly technological nations reached a point of crisis somewhere in the quarter of a century that followed World War II. These societies plainly are in some kind of transition. The value systems that ultimately inform decision-making obviously are being reordered.

All this is bound to affect, directly and indirectly, the course of modernization in the rest of the world. The unanswered question is whether the transition to postindustrial society will be purposeful and relatively painless, permitting the international system to adjust by stages, or whether it will involve national disasters with traumatic impacts on the world at large.

The disaster alternative clearly cannot be ruled out. One need only glance at the energy problem in the United States. The industrial plant, the transport system and the very patterns of habitation in this country are all based on the assumed availability of abundant, cheap energy. Yet in 1979, after five years of circular debate about national energy policy, in the face of constantly escalating energy prices, there was still little indication of awareness at the Federal level that the end of a very brief historical period of cheap and abundant energy requires pervasive redevelopment of industrial society.

It will make a very large difference—not only to Americans but to the rest of the world as well—whether such redevelopment can proceed within deliberate stages or whether it must await the shock of galvanizing disaster.

3

Rich and Poor:
The North-South Dimension

It was self-evident at the time that the year 1945 would loom large in the history books. The first atomic explosion introduced a weapon whose unheard-of destructive force promptly brought an end to World War II—and threatened an end to civilization. And in that same year, there appeared above the ruins two superpowers— both soon to be massively armed with this new weapon—as the new centers of world politics.

In retrospect the year acquires even greater historical significance. For at the end of the war two other trends appeared, little noticed at first, that would soon converge to transform the map of half the world and impinge powerfully upon the politics of the final quarter of the 20th century.

First, the exhaustion of the European combatants ensured the success of the national independence movement as the European colonial system began to unravel in Asia, the Middle East, Africa and the Caribbean. Within a few decades, some 80 nations, inhabited today by about one-third of the world's population, became independent. It was possibly the most massive transfer of political power in such a short time in the history of the world. Henceforth, if their peoples were to suffer tyranny, it would be at the hands of their own nationals; if mistakes were to be made, it would be their own mistakes. Membership in the UN quickly provided the new nations with a place in the sun, a chance to be heard, a license to participate in world affairs.

Second, the rate of technological innovation increased so rapidly after 1945 that the phenomenon soon was referred to as the knowledge explosion. This, too, was partly stimulated by World War II. In the United States, during the war years, a combination of heavy demands for food, plus pressures on manpower supplies for the armed services, had greatly accelerated the industrialization of American agriculture. Military needs had also stimulated the development of industrial products and processes—notably electronics and the beginning of the computer revolution—that later would become leading peacetime industries. Thus energized and redeployed to meet pent-up consumer demand, the productive capacity of the U.S. economy doubled in just over two decades of steady growth. A profusion of new products tumbled into the marketplace. New industries—television, plastics, electronics and others—came into being and expanded radically. There was a jump in the rate of technological obsolescence. And perhaps most significantly for the prolongation of the knowledge explosion, the U.S. government had become in wartime, and remained in peacetime, a heavy financial supporter of basic research and technological development.

Decolonization and Technology

It did not take long for these two new trends to meet and interact. The knowledge explosion further encouraged the leaders of the newly independent nations to embark on the path of modernization, to build their political platforms around plans for "development," to favor the growth of industry in the allocation of resources, and to give top priority to prompt acquisition of the most modern technology. Steel mills and international airlines were among the leading symbols of modernization at work in the developing world. In India, the giant among developing countries, Mahatma Gandhi's vision of a nation committed to cottage industry and the simple village life did not survive his death in 1948. The simple villages survive, a stubborn reality; but the aim of every Indian government from that day to this has been modernization.

No longer, then, after 1945, was the idea of progress through science and technology an essentially Western phenomenon. Now this idea was loose in the world at large—where its clash with fatalism and theological and cultural tradition was potentially more violent than it had been in the West during the early Industrial Revolution. The fever of modernization had spread to all the continents.

At first it was transmitted, outside of political channels, through the wartime spread of the products and skills of the industrial era—radio, aircraft, motor vehicles, antibiotics—and through post-war trade. Even the rulers of emerging new nations, caught up in the sheer euphoria of independence, gave little thought to the enormous economic and technological needs of a modern state.

It was not many years after the first great wave of decolonization, however, that the euphoria gave way to second thoughts. Political independence had not ended the economic, trade, communications and technological dependencies of the former colonies upon the former overlords. Much of what economic development had occurred during the colonial era was designed to complement the domestic economies of the colonial powers. Transport systems and cities were built to serve export markets. Modernization began in selected sectors of the colonial economies, while other sectors were left to stagnate in their traditional condition. The whole international economic order had been designed by the industrial nations to serve their own needs—or so it seemed to many in the newly independent countries.

Even in the early postwar years, these facts were not entirely ignored in Washington. As early as 1949, President Harry S. Truman had recognized the needs of developing countries by launching his small, landmark "Point Four" program of technical assistance. Through the 1950s, the United States and some European countries gradually elaborated broader programs of financial assistance for development by governments, or by the World Bank and other international lending institutions which Northern governments supported. These efforts supplemented the much larger profit-oriented flow of capital and technology by private investors in the developing South.

Thus, what would soon come to be called the politics of development, or North-South relations, was well begun by 1961 when President John F. Kennedy proposed in the UN a "Decade of Development." Ten years seemed a long time then, but today the third Development Decade is under debate in the UN, and there is no end in sight. Moreover, the complex ramifications of the development process, and of North-South relations, became clearer as time passed. Not the Southward flow of capital alone but also the opportunity to trade with the North on more advantageous terms became central to the development credo of the new nations—and the industrial nations began to respond with trade preferences and with arrangements to stabilize international commodity prices. The wide range of "infrastructure" needs in developing nations— schools, roads, communications, airports and seaports, weather bureaus, hospitals, agriculture and industrial technologies, and training in skills of all kinds—became the substance of thousands of development projects and plans, and called into service the whole array of sophisticated skills in the UN Specialized Agencies. More or less coherent national development plans began to appear. Tens of thousands of students from the developing world came to study in Northern universities.

Numerous but politically and economically weak, the developing nations learned early to increase their influence with the North by joining forces in a common negotiating front. A notable product of their alliance is the UN Conference on Trade and Development (UNCTAD), first convened in 1964 and since institutionalized as a continuing global forum for airing the claims and concerting the positions of the developing countries on North-South economic issues. The solidarity of the "third world" on development issues thus became a major political fact—and sometimes almost an end in itself from the viewpoint of developing countries. The Group of 77, formed at the first UNCTAD as the lobbying arm of the developing nations, survives as a permanent part of the UN's machinery. Its membership is now about 120.

Science and Technology on the Agenda

Such was the political context in which the application of science and technology to development first emerged as a distinct item on the international agenda. In 1963 the UN convened an international conference of scientists on the subject. An international advisory committee of distinguished scientists was created, and in 1971 it recommended—to largely inattentive governments at first— a "world plan of action" for technological research into medical, nutritional and other problems of importance especially to tropical developing nations. A stream of committee reports and resolutions on the subject began to expand the files of the UN as the international community wrestled with a scientific-political-organizational riddle on which few specialists and still fewer politicians could agree. But the good news was that the international community had at least identified a question of capital importance: How can science and technology serve the development of nations and the conquest of poverty? Sorting out the multiple answers would be a work of many years.

By the end of the 1960s, measured by traditional criteria, economic growth in the developing world was proceeding unevenly but fairly fast. In an address devoted to North-South relations in March 1979, Secretary of State Cyrus Vance reported that for more than a quarter of a century, per capita income in the developing countries as a whole rose by nearly 3 percent annually—about 50

Attending opening of May 1979 UNCTAD V conference in Manila were (l. to r.) Carlos P. Romulo, president of the conference; UN Secretary General Kurt Waldheim; President Ferdinand Marcos of the Philippines; and Mrs. Marcos.

United Nations/Photo
by M. Bolotsky

percent better than historical growth rates in Western nations during their period of industrialization. Life expectancy shot up from 42 to over 50 within two decades, an improvement which had been spread over a century in the industrial nations. Adult literacy went from one-third to one-half of the population of the developing countries.

By such standards, the modernization process appeared to be well launched. But then came the oil-exporting countries' historic price coup of 1973. This gave the countries of the South potent new leverage and promptly encouraged new and far broader demands by all developing countries, oil-rich and oil-poor alike. No longer were they content simply to call for more development aid or for a few trading preferences. Their target now was nothing less than the existing international economic system, which they had had no hand in shaping and which they believed to be inherently biased in favor of the rich, industrialized nations. Simply to join that system and become part of it, they were now saying, could only perpetuate dependencies from which they wanted to escape. So they proposed to change the system—to "restructure" it so as to eliminate what they saw as its inequities. And they wanted a full voice in what they proclaimed as the New International Economic Order.

But economic systems are not restructured by proclamation. Not only could the Group of 77 not gain Northern assent to their main demands; they could not offer a coherent description of what the new order should look like. Herein lay their dilemma: their negotiating strength depended upon maintaining strict solidarity, but they could not agree among themselves on anything more than largely abstract generalities. The notion of solidarity, of course, rests on the myth that "developing countries" are in a category of nations with consonant interests, when, in fact, they are extremely disparate from every point of view: huge and tiny, resource-rich and resource-poor, with many or few educated people, democratic or authoritarian, allied with West or East or neither. Little wonder that, in the course of preparations for UNCSTD, one third-world representative complained that the Group of 77 suffers from "a horror of specificity."

In any event, the inability of the nations of the South to formulate concrete demands in negotiable form made it easy for the representatives of the North to evade any serious discussion of international economic reforms. Yet there was irony in the Northern position too, for their spokesmen were defending an international economic system which they manifestly no longer knew how to control and which plainly was working poorly from the viewpoints of developed and developing countries alike. A demand for a new order might have come almost as logically from the industrial North as from the developing South.

Poverty remains a challenge to science and technology and a controversial issue in the North-South dialogue.

United Nations Fund for Population Activities (by Mark Edwards)

The Developing World's Manifesto

It was in such an atmosphere that in early 1974 the developing countries pulled together their demands, codified them as best they could, and laid them on the table in the main forum of the world community—the General Assembly of the UN, convened in its Sixth Special Session to debate the New International Economic Order.

The late Colonel Houari Boumediene, Algerian president and spokesman for the Group of 77, opened the debate. His posture was one of confrontation. He missed no chance to score a debating point, to rub salt in wounds, to spot imperialist and neoimperialist plots and conspiracies. Neither Boumediene's speech, nor those that followed, nor the rhetoric of the final resolutions proclaiming the new order, did much to clarify the substantive content of an alternative international economic system. But the session's political message was clear.

Decolonization had ended an age of empire, but had opened an age of world politics in which there are many more actors with the determination—and increasingly, the ability—to make themselves heard on the agenda and organization of the postcolonial world. At the top of their priorities is the attainment of what they see as a more equitable restructuring of the world economic system. For them this is not a technical but a political question—the next stage of the independence movement, whose accomplishment is essential to their further economic progress. Thus is the politics of modernization linked to the politics of independence.

The North-South confrontation had appeared to reach a dangerous pitch at that Sixth Special Session. There was an apparent turnaround the next year at the Seventh Special Session, when the United States took a strikingly more forthcoming line. There followed a faltering and largely unsuccessful 18-month effort, ending in 1977, to launch a serious North-South dialogue at the Conference of International Economic Cooperation in Paris. At this writing the short-term trend remains unclear. The South still seems somewhat paralyzed by its diverse members' perceived need for solidarity, while the North seems preoccupied with traditional security issues, with efforts to patch up the existing economic system, and with the narrow technicalities rather than the broad politics of international economic life. Thus the danger persists that North and South could drift increasingly into a dialogue of the deaf.

This is the relevant political context of the world agenda item called science and technology for development.

4

Development Strategy Transformed

Ideas about science, technology and development were in flux in the developing world as well as in the industrial world in the 1970s. When the fever of modernization began spreading in the wake of decolonization and the knowledge explosion, it was widely supposed that industrialization in the former colonial world would be more or less a replication of the story of industrialization in Western Europe and North America. Some new nations, to be sure, drew part of their inspiration from one-party, centrally planned states, such as the Soviet Union, China or Yugoslavia; but various mixes of state and private enterprise—much of the latter being foreign—were more characteristic. Technology thus flowed from the industrial North through both private and governmental channels.

The early plans and strategies adopted by the developing countries generally reflected the cluster of ideas surrounding the

Western notion of progress. Rapid overall economic growth was the main target. GNP and per capita income data were the main yardsticks. And the industrial sector was expected to lead the way to modernity.

The strategy agreed upon for the first UN Decade of Development, adopted in the early 1960s, was squarely within this mold. The overriding goal—which was very nearly achieved as a worldwide average—was to reach and sustain a 5 percent annual rate of economic growth throughout the developing world.

By the 1970s the more reflective leaders in the developing world were having second thoughts. Despite unprecedented rates of growth by traditional standards of measure, modernization was not delivering the expected gains in social well-being, or national strength and coherence, let alone in personal happiness. Population growth was eating up most of the production gains. World food production in particular barely stayed ahead of the world population boom, and in many countries it fell seriously behind.

For another thing, the benefits of development were not trickling down to those on the bottom rungs of the economic ladder. Even where national statistics showed more income, more food, more schools, more housing, the number of very poor, hungry, illiterate, homeless and unemployed people was greater than ever—and growing. The gap between the affluent modernizing sectors and the stagnating, often absolutely poor sectors, was becoming more apparent—and more embarrassing to political leaders. Modernization was worldwide but shallow—mainly affecting the glossy surface of society.

New Facts, New Concepts

More fundamentally, the notion that Europe, America, the Soviet Union, China or any other society could provide a "model" of development was beginning to come unstuck in the developing areas—even as the original idea of progress was beginning to come unstuck within the Western industrialized countries. New ideas were being explored on both sides of the North-South divide, ideas which showed a steadily growing insight into both the process and the purposes of development, whether in highly industrialized or newly developing economies.

▶ The sudden awareness of environmental constraints on indiscriminate economic growth, which led to the Stockholm Conference in 1972 (see next chapter), soon gave birth to the concept of "eco-development"—meaning modernization strategy shaped to accord with the limitations of natural systems and resources.

▶ In 1973 the Arab oil embargo shattered the assumption in the industrial world that energy supplies could be taken for granted. By drawing attention to the finiteness of fossil fuel, it also foretold a time when "modern" would mean less, not more, dependence on traditional energy sources—no matter how hard official and public opinion would try to ignore the message.

▶ In 1974 the World Population Conference was held in Bucharest in August, and less than three months later the World Food Conference was held in Rome. When they were over, it was harder than ever to pretend that the population problem and the hunger problem were not intricately linked. It also was harder to believe that industrialization could produce tolerable societies unless the food and population problems were resolved.

▶ More and more Northern governments and international institutions, notably the World Bank, were insisting that modernization policy must include a direct attack on the roots of poverty. This insistence stemmed from the growing perception that the vicious circle of poverty-hunger-population-unemployment becomes self-perpetuating unless there is a direct intervention to help meet the basic needs of the poorest of the poor. The World Bank, the U.S. bilateral AID program, and some of the UN Specialized Agencies proceeded to reshape their program priorities in the second half of the 1970s for a more concentrated attack on the base of poverty as an integral if not predominant aspect of the modernization process. Perhaps the most promising approach to this task is the movement to organize basic services at the village and neighborhood levels with local people largely in command of both planning and operations.

The World Health Organization, the UN Children's Fund (UNI-CEF), and some private agencies are forerunners in sponsorship of highly decentralized, basic service delivery systems to provide simple health care, family planning, nutritional advice and child care at the local level.

▶ Developing countries themselves have increasingly espoused the idea that modernization should be pursued more and more through the self-reliance of developing countries, individually or as a group. In a negative sense, this is a reaction against the dependence of developing countries on Northern industrial countries and has led to talk about "selective de-coupling" of economic relations between North and South. In a positive sense, though, it opens potentially significant prospects of assistance from the "middle-income" countries with relatively well-developed technical resources, like Brazil, Mexico, India, Israel and others, to countries in very early stages of modernization, with a corresponding relaxation of demands upon the industrialized nations.

Some Emerging Issues

These are indications of a general ferment in contemporary thinking about the development process as governments probe for viable paths toward modernization. Out of this ferment and the running debates it has engendered, a number of issues have emerged that bear directly on the problems of using science and technology for development.

Appropriate technology. One might expect easy agreement on the proposition that any technology selected for any task should be appropriate to its purposes. But it is not that simple.

Many development experts have stressed the point that much of the most sophisticated technology has been designed to minimize the use of manpower, while technologies appropriate to conditions in developing countries are likely to be labor-intensive to take advantage of that most abundant resource of the developing world. Appropriate technologies also would be those which make maximum use of local materials. They should serve to strengthen rural and village life, and diminish pressure on overburdened cities. And they should be within the managerial and technical capabilities of the society in question.

These points make sense to the experts, but they have an ominous ring to many political leaders of developing countries. Their ambitions tend to be focused on obtaining the latest and the best in the way of technology. Anything short of that appears second-class, perhaps obsolete or obsolescent. Indeed, some developing country spokesmen have denounced Western support of "appropriate" or "intermediate" technology as a conscious effort to keep third-world countries in a permanently inferior state of technological development. Ironically, Northern advocates of appropriate technology may be finding a more receptive audience among the critics of industrialism in their own countries than they do in the developing world. This issue has produced much confused debate and remains an obstacle to North-South understanding on strategies for technology transfer.

Basic human needs. The notion that development programs should give first priority to meeting the basic needs of all people for food, health, literacy and shelter also has become a contentious issue in North-South affairs. The "basic needs" development doctrine stems from conviction that direct intervention is needed to break the vicious circle of absolute poverty. Its roots can be found in many documents drafted by experts from the third world.

This bottom-up approach to development was welcomed with enthusiasm in some of the industrialized countries, notably the United States. Political support for foreign aid had become increasingly difficult to muster, especially as evidence accumulated that the benefits of development were going overwhelmingly to those already in relatively favored circumstances. A "basic needs strategy" seemed to offer the most persuasive argument in favor of continued development assistance from the affluent countries.

As in the case of "appropriate technology," however, Western advocacy of the basic needs approach aroused suspicions and resentment among some in the developing countries. Was this, they

The "basic human needs" development program emphasizes "bottom-up" assistance, such as medical care.

asked, an effort to convert the great drive toward modernization into a global welfare program to keep developing countries from acquiring modern industrial systems for the indefinite future? What's more, developing country leaders know that a serious attack on the problems of the poorest of the poor implies heroic measures of internal reform—like more equitable land tenure systems and access to credit—which raise grave political dangers for regimes still struggling to consolidate their power base and to attain a greater measure of stability at home.

Despite such difficulties, the basic needs approach has helped to focus attention on the general failure of development programs—with conspicuous exceptions in the cases of Taiwan, Korea, Singapore, Hong Kong and a few other countries—to reach the problem of absolute poverty. It also has helped to raise priorities for agricultural and rural development, especially in early stages of the modernization process. As a general policy or strategy for development, however, there is little consensus and much contention in the ongoing debates.

Multinational corporations. The role of multinational corporations (MNCs)—or transnational corporations (TNCs) as they are now called by UN agencies—probably has inspired more emotional rhetoric than any other aspect of the technology-for-development issue. It also illustrates the potentially dangerous contrast that often exists between what political leaders say on conspicuous public platforms and how they behave in private.

Unquestionably, the modern multinational private corporation operates with unrivaled effectiveness in an economically interdependent world. Nor does there seem to be much argument against the proposition that these corporations have been the major conduits for the transfer of new technologies to developing countries—including, perhaps most importantly, the managerial and administrative components. This, no doubt, helps account for the fact that many developing countries, even as they publicly denounce TNCs as agents of neocolonialism, work pragmatically with such corporations to help carry out their modernization programs.

The basic issue, however, is when and under what circumstances the profit motives of a private corporation can be accommodated with the best interests of a developing country. Countless arguments have been published on all sides of that question without lowering the decibel level of political debate. And it is much to be doubted whether any general answer at an abstract level—such as the proposed UN "code of conduct" to govern the operations of corporations doing business overseas—can ever fit all cases or command general assent. A more fruitful approach might be to seek instruction from the study of a broad range of case histories—some of them "horror stories" of exploitation on one side, arbitrary expropriation on the other side, or both; some recounting ventures deemed highly successful for both investor and host country; and most falling in the mid-range between best and worst. In any case, it is predictable that the MNCs will continue as vitally important participants in the world system, and that the controversy over their role will continue to rage.

Intellectual property. From the viewpoint of developing country leaders, the Paris convention of 1883 protecting inventions, trademarks and industrial designs is a major obstacle to a more equitable world economic order. This is a limited case of a much broader issue that seems likely to surface and intensify as time goes on: the concept of knowledge as the heritage of all mankind.

The notion that human knowledge should be held in common as a free commodity available to all appears, at first blush, to be an extremely attractive moral, philosophical and political proposition. As things stand now, there are many inhibitions against free access to information, including classified government data, proprietary information, and commercial secrets. The knowledge embedded in patented technologies is generally seen by private companies as a product of their own expensive research and development that has been bought and paid for and that is available to others only for what is considered adequate compensation. On the other hand, technologies in many important areas, like agriculture and public health, are overwhelmingly in the public domain and yet remain grossly underutilized today.

The modernization experience, however, is throwing a new light on information as a developmental resource with inherent characteristics quite unlike the characteristics of finite material resources. In a basic sense, information is now seen as an underlying component of all other resources and as a precondition for the successful use of any kind of technology. This evolution in the perception of the role of information in the modernization process comes at a time of turbulent changes in technologies for gathering, storing, processing, communicating and applying information. The industrial society, some people believe, already is being replaced by an "information society." And this issue is emerging on the global level: What restrictions, if any, should limit access to the sum of knowledge accumulated to date by the human race?

The foregoing contentious issues have been prominent in the course of preparations for UNCSTD. No doubt, they will be debated for years to come—though hopefully in less simplistic formulations. All of them belong to the baffling class of problems which are partly technical or economic in nature but are also partly political, even philosophical and ethical. For such questions it is extremely difficult to arrive at answers that are both clear and widely accepted.

Small wonder that harried diplomats have difficulties in enunciating clear-cut and comprehensive policies for science, technology and development that sometimes are demanded of them!

5

Planetary Politics in
Its Infancy

In earlier chapters we noted the spread of the global fever called modernization; the new demands of the developing South upon the industrial North; and the need to redefine words like progress and modernization so that the power of science and technology may more consistently serve human values. UNCSTD, convening in August 1979, reflects all these historical trends.

But the conference also has another significance. Through it, the problems of science and technology take their place among a growing class of problems and needs that are intrinsically global— aspects of a new interdependence among nations in a crowded and rapidly modernizing world. In response to this new class of problems, the nations have devised new methods of cooperation and negotiation through a series of world conferences held under the aegis of the United Nations. It is the dawn of an age of planetary politics.

World Conferences of the 1970s

The first of these conferences was held in Stockholm in June 1972: the UN Conference on the Human Environment. In the years that followed, UN conferences were assembled on a wide range of other problems of interdependence. Notable among them have been conferences on world food (Rome, 1974), world population (Bucharest, 1974), the role of women (Mexico City, 1975), the human habitat, both urban and rural (Vancouver, 1976), global water resources (Mar del Plata, Argentina, 1977), desert encroachment (Nairobi, 1978), employment (Geneva, 1976), primary health care (Alma-Ata, U.S.S.R., 1978) and technical cooperation among developing countries (Buenos Aires, 1978). Many of these subjects are only recently recognized as having global significance. They have taken their place alongside long-standing problems of interdependence, such as trade, development, and that most ramified of all world legal problems, the effort to write a new world law of the sea and the seabed.

That there will be more such UN conferences is a certainty. Among those in the planning stage as this is written is a world energy conference, scheduled to convene in 1981.

Much earlier than the 1970s, to be sure, many international discussions of such subjects had taken place—but primarily among expert technicians and specialists. The crucial fact about the conferences of the 1970s is the level at which they have been held. All were conferences of official delegations with authority to take positions on behalf of governments. This was the real achievement of the much-maligned UN: to get the governments of the world to start talking to each other at the political level about some of the most important and pressing problems of our times.

Several points need to be made about why such an innovation was necessary, and how and why it works.

In part, the need for something new arose because the new problems did not fit the old organizational molds. The technical talents available to the international system are lodged primarily in the family of UN specialized agencies, each with its corps of experts in such fields as tropical medicine, plant genetics, meteorology, educational psychology, etc. Environmental, resource and technological problems often cut across all these disciplines and a good many more. Considered broadly, so do the problems of food, population, water resources, energy, etc. The solution has been to organize UN conferences outside the framework of the specialized agencies, with secretariats specially recruited for the purpose. Each conference thus has freedom to draw on the special expertise of many agencies without becoming the bureaucratic prisoner of any.

World conferences such as the UN Conference on the Human Environment (Stockholm, 1972) have stirred public awareness of global issues. This Swedish environmental group was protesting Brazil's policies in the forest state of Mato Grosso.

United Nations

Still more importantly, the need arose from the simple fact that the problems on this new world agenda are manifestly beyond the capacity of any single nation, or even group of nations, to cope with. Either they are inherently global in the physical sense—like the impact of the burning of fossil fuels on the world climate, or the vulnerability of the earth's protective ozone layer; or they are of such widespread concern that their economic and human effects cannot be contained by national borders, and no government dares to ignore them—like population growth and the spread of hunger. They are, in fact, not so much *inter-national* as *intra-human* problems—affecting the human family as a whole, and requiring cooperative solutions at the global level. Moreover, in many cases the problems are so novel that political decision-makers have only begun to become seriously aware of them in very recent years. The family members, therefore, have no rational choice, as residents of the same planet, but to study the new problems together and cooperate in dealing with them.

Educating National Governments

It is the world's 160 or so national governments, when all is said and done, that must do the studying and the cooperating. Whatever their limitations, it is governments that possess the power to act—to mobilize money and talent, to generate and control action on a large scale and over long periods of time. Yet the reaction time of governments, especially on new problems, can be painfully slow. Recognizing these basic facts, the organizers of world conferences of the type begun at Stockholm made their plans accordingly. They provided for a preparatory period of as much as four years before each conference—time for governments to educate themselves on the subject in hand. They took care to assure that the conference itself would be attended by representatives from each country with enough rank to influence necessary decisions by their political leaders at home. And they contemplated that each such conference would have a vigorous afterlife. Each has given rise to a follow-up plan of action and some kind of institutional arrangements—in some cases a new international organization or agency—to carry on the work. Each conference is thus a high point in a process that begins years earlier and may continue for many years after. Each conference topic, be it environmental protection, food, population planning, or water resources, becomes one of the staple topics of planetary politics—an agenda for ongoing work on problems that cannot be solved but only managed.

Good preparation is particularly vital in engaging the efforts of national governments. Governments are notoriously adept at putting off problems that are unfamiliar and may be politically difficult to face. But a government that has promised to show up at a formal conference on a specific date, in a certain place, prepared to state and defend publicly an official position on the announced subject, has swallowed a strong antidote to bureaucratic lethargy. It has no choice but to get on with the necessary homework. The point can be illustrated by reference to experience with the earliest and the latest of these conferences.

About a year before the Stockholm conference, a representative of the UN Secretariat toured a large group of nations in the Middle East to inquire into the state of their preparations. He was received by prime ministers and foreign ministers. Some of them turned out to be unaware that the conference was to be held at all. Most of them did not know their government was expected to contribute a paper. And in no case had any official in the government been assigned responsibility for conference preparations. Five years after the Stockholm conference, a survey found that 70-odd governments had established official agencies, many at the cabinet level, to deal with problems of the environment. In many of these countries, first awareness of the existence of environmental problems had been stimulated by the preparations for the Stockholm conference.

Nor is this sort of impact limited to developing countries. Indeed, it can be highly significant in the most advanced industrial nations, including the United States. Preparations by the U.S. government for UNCSTD illuminate the point.

The United States, by any measure, has had more experience

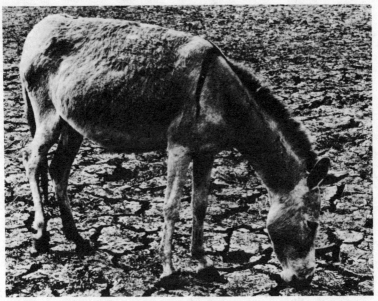

**Desertification, one of the results of misused technology, was the subject of a
UN conference in 1978.**

than any other nation in the application of technology to the
international process of modernization. Yet, up to 1978, no U.S.
policy on the subject had ever been formulated, nor, for that matter,
was there any place within the executive branch with a mandate
even to think about the problem. In the midst of the preparations for
the UN conference, a publication of the United Nations Association
of the United States of America cited the relationship between
technology and diplomacy as "the most overlooked problem facing
U.S. foreign policy today."

As this was written in spring 1979, the agony of preparations for
Vienna was still unfinished in the Washington bureaucracy; but at

least it could no longer be said that the problems of science and technology in development were being "overlooked" by U.S. officialdom. The government had before it, among other preparatory documents, a specially commissioned report by the National Academy of Sciences identifying priority areas of research of special interest to the economies of developing countries. Twenty-two such areas were identified in the field of agriculture alone. Under most active policy consideration was a plan first announced by President Jimmy Carter to create a U.S. institute for international technological cooperation, whose functions would include "helping developing countries to assess, select, invent . . . and adapt technologies; to strengthen their supporting science base; and to improve their science and technology infrastructure." Such plans showed an important evolution in U.S. government thinking and planning on this subject, brought about directly by the need to prepare for the Vienna conference.

National governments cannot be expected, of course, to thank the international community for pressuring them into working on problems they should be working on anyway. But this is a significant effect of the unheralded arrival of the age of planetary politics: The external pressures of the new global agenda have begun to force national governments to put their internal houses in order and begin to come to terms with emergent global issues.

The world conference process of articulating global issues leaves much to be desired. At best these conferences are messy affairs. A gathering of 150-odd delegations from sovereign, disparate and quarrelsome governments is bound to produce a large amount of political posturing, a high level of boring ideological rhetoric, and a long list of proposals that are irrelevant to the agenda and impossible of attainment. But the often stormy environment of the plenary sessions of these conferences—where media coverage is almost exclusively focused—can be deceptive. As in any parliamentary body, most of the serious work is done in committees. And the real bargaining, if and when they get around to that, is done within small groups safely out of public sight.

Some of the world conferences may have accomplished no more than a heightened awareness of a new global problem—though that in itself may be the beginning of wisdom. Others have led to specific cooperative agreements that otherwise would not have come about, such as the program to control pollution in the Mediterranean Sea, sponsored by the UN Environment Program. And one—the World Food Conference—has brought the international community closer to an adequate world plan of action than has been the case in any other major world problem area.

When negotiations in planetary politics do not succeed, the participants do not go to war. They do not even stop talking to each other. They repair to their drawing boards and try again in the next forum, and, if need be, in the next. So far, at least, no nation has succeeded—though some have tried—in preventing a debate on global issues; nobody has walked out of a world conference;* nobody wants to take responsibility for breaking off the dialogue.

This is the broad political setting in which nations are attempting to come to grips with the fevers of modernization and the role of science and technology in that worldwide experience. It is a far cry from a constitution-writer's dream of world government—or even from the image of a "Parliament of Man" operating under *Robert's Rules of Order.* Up close, the business of planetary politics looks like a somewhat disorderly gathering of the tribes. And perhaps that's just about where we are in the evolution of a world society.

---

*The Soviet Union, for political reasons long since irrelevant, boycotted the Stockholm conference in 1972, but is an active participant in the UN Environment Program to which the conference gave rise.

6

Science, Technology and Politics: Managing Modernization

It could not be known as this was written how far the UN's Vienna conference of 1979 would succeed in bridging the gaps of understanding and purpose between North and South in regard to the uses of science and technology in development. Inevitably, this session would be at best one step in a process extending far into the 21st century. There are no short cuts, for the precious developmental assets we call science and technology involve not only hardware, which can be quickly transported and installed, but nonphysical resources which cannot: educational systems, research facilities, professional societies, library services, museums, information systems. And they also involve, at every stage, an enormous range of skills—from the skills of politics and management to the simple ability to operate and maintain machines. Moreover, technology interacts so intricately with the circumstances and culture of each country that it would be impossible for all who seek its benefits to follow a standard "model." Rather, it must be carefully adapted to the particular needs and capacities of the user. Hence, developing countries cannot be mere passive beneficiaries of imported technology; rather, the process must be a joint enterprise in which user and provider collaborate from the beginning.

And for the same reason, no matter how various the needs of different countries, one indispensable component of any successful development plan must be a serious, systematic, long-term program to strengthen indigenous scientific and technological capabilities in every country and region. To ignore this need, and to rely solely upon the importation or transfer of ready-made technology, would be to build upon sand—indeed, to perpetuate the very dependence that many advocates of technology transfer most eloquently oppose. Any technology, even the most sophisticated, will sooner or later become obsolete; but what never becomes obsolete, once it is acquired, is the intellectual and managerial capacity to develop new technologies to meet newly felt needs.

Modernization—the Long View

The management of modernization, then, is an overarching task for the rest of this century and beyond—one that inescapably will intensify the relations between science, technology and politics.

For it is inconceivable that man would relent in his pursuit of scientific knowledge. What is needed rather is a clearer understanding of what kinds of questions science can be asked to answer and what must be left to the realms of values, ethics and subjective choice.

It is just as unthinkable that man would cease to search for improved technologies to lighten the burdens of prevailing misery and to enhance his environment. What is needed is a more mindful matching of technological potentials with pluralistic social goals.

Meanwhile, the politics of modernization burdens each national society with responsibility for deciding "what unavoidably must be changed and what must at all costs be preserved" as societies probe for that delicate balance between the universal urge to change society for the better and the universal need for some measure of social and cultural continuity.

At the international level, the application of science and technology to development is a subject that will not go away after a world conference has been held, regardless of the level or lack of accomplishment at that one session. It is certain to remain indefinitely on the agenda of global issues.

Government and Science in Partnership

And—a fact of great potential importance—interest in this subject will not be restricted to governments. We know this because, parallel to each world conference, there has come into being a network of nongovernmental organizations—publishing their own preparatory studies, holding their own unofficial conferences, pressing their views on governmental agencies, looking over the shoulders of official delegations.

This process began at the Stockholm conference where environmental activists conspicuously competed with the official conference for public attention. By the time of the Vienna conference, a series of preparatory nongovernmental conferences on science, technology and development already had been held. Some were organized by citizen groups concerned with broad issues of public policy. Others were meetings of scientific bodies concerned with clearly defined technical questions. One of the most fruitful sessions combined both these characteristics. It met early in 1979 in Singapore—an extraordinary consortium of 19 international scientific bodies representing the physical sciences, the life sciences, the social sciences, engineering and technological professions. These rather disparate sponsoring groups agreed to continue working together in support of the modernization of developing countries, and to establish nongovernmental networks within the international scientific community for that purpose. They also agreed upon the central importance of strengthening the scientific and technological resources of the developing countries.

Thus, as the first decade of planetary politics was coming to a

The proper management of
science and technology has
become a global concern.

United Nations

close, it appeared that some important barriers were coming down—barriers that had obstructed communication between the natural and social sciences and the technological disciplines, that drew rigid lines between governmental and nongovernmental functions and, most unfortunate of all, that separated the international scientific community from the international political process.

It is none too soon. For modernization is now irrevocably in the mainstream of world politics. The rate of change is rapid, and politicians in every continent are daily called upon to decide what is to be changed and what is to be preserved. Never has it been so necessary for the management of that turbulent process, whose effects reach beyond the borders of every nation, to combine the political power and skills of government with the light of science.

Talking It Over

A Note for Students and Discussion Groups

This pamphlet, like its predecessors in the HEADLINE Series, is published for every serious reader, specialized or not, who takes an interest in the subject. Many of our readers will be in classrooms, seminars or community discussion groups. Particularly with them in mind, we present below some discussion questions—suggested as a starting point only—and references for further reading.

Discussion Questions

Modern technology has now spread to every nation in the world. What are some of the resultant benefits and problems?

Scientific discovery and technological innovation have encountered resistance for hundreds of years. Discuss the four main sources of opposition. Do you agree or disagree with the basis for such opposition?

Why did anthropologists and others begin to question whether science and technology necessarily contribute to "progress"? Has the worldwide advance of technology been helpful or harmful to human values?

In 1970 the United States enacted the National Environmental Policy Act. What are some of the implications of this legislation? Do you believe the government should choose the quality-of-life problems (health, environment, etc.) that technology should solve? Or should such decisions be left to the business and scientific communities? Give reasons.

What does the author mean when he refers to the knowledge explosion? Where had it led?

The UN is now in its third Development Decade. What have been some of its accomplishments and shortcomings? Do you think the UN Conference on Trade and Development has been successful in its efforts to help the developing nations? Why or why not?

Contrast the politics of modernization in the third world with the earlier politics of modernization in Europe and North America.

What are some of the new concepts of development that are currently being discussed and explored? What are appropriate technologies for developing countries?

Discuss the role of multinational corporations in the transfer of technology to developing countries. Have they done a useful job? What are some of the criticisms leveled at multinational corporations?

What does the author mean by planetary politics? Have the UN conferences on the environment, food, population, etc. helped to solve global problems? What are some of the things you think could be achieved at UNCSTD?

READING REFERENCES

Baranson, Jack, *Technology and the Multinationals.* Lexington, Mass., Heath (Lexington Books), 1978.

Basiuk, Victor, "Technology and World Power." *HEADLINE* Series 200. New York, Foreign Policy Association, April 1970.

Black, Cyril E., *The Dynamics of Modernization—A Study in Comparative History.* New York, Harper & Row, 1977 (paper).

Boorstin, Daniel J., *The Republic of Technology*. New York, Harper & Row, 1978.

Cahn, Robert, *Footprints on the Planet: A Search for an Environmental Ethic*. New York, Universe Books, 1978.

Cleveland, Harlan, and Wilson, Thomas W., Jr., *Humangrowth*. New York, Aspen Institute for Humanistic Studies, 1979.

Goulet, Denis A., *The Uncertain Promise—Value Conflicts in Technology Transfer*. Washington, D.C., Overseas Development Council, 1977.

Jones, Graham, *The Role of Science and Technology in Developing Countries*. London, Oxford University Press for the International Council of Scientific Unions, 1971.

"Limits of Scientific Inquiry," *Daedalus*. Spring 1978. Cambridge, Mass., American Academy of Arts and Sciences.

Marx, Leo, *The Machine in the Garden: Technology and the Pastoral Ideal in America*. New York, Oxford University Press, 1977 (paper).

Mead, Margaret, ed., *Cultural Patterns and Technical Change*. New York, New American Library, 1955.

Mobilizing Technology for World Development. Report of the Jamaica Symposium. Washington, D.C., International Institute for Environment and Development, 1979.

Muller, Herbert J., *The Children of Frankenstein—A Primer on Modern Technology and Human Values*. Bloomington, Ind., Indiana University Press, 1970.

National Research Council, *U.S. Science and Technology for Development—A Contribution to the 1979 UN Conference*. Washington, D.C., USGPO, 1979.

Science and Technology for Development. U.S. national paper prepared for the 1979 UN Conference on Science and Technology for Development by the Office of the Coordinator of that conference in the Department of State. (International Organization and Conference Series 139.) Department of State Publication 8966, 1979.

Technology In Society, Spring 1979. New York, Brooklyn Polytechnic Institute.

The Smithsonian Book of Inventions. Washington, D.C., Smithsonian Institution, 1978.

Tolley, George S., ed., "International Science & Technology; The Policy Gap." Chicago, Ill., The Chicago Council on Foreign Relations, 1979.

White, A.D., *A History of the Warfare of Science with Theology in Christendom,* vols. I and II. New York, Dover Publications, 1960.
Winner, Langdon, *Autonomous Technology: Technics-Out-of-Control As a Theme in Political Thought.* Cambridge, Mass., MIT Press, 1977.

Our Next Issue, HEADLINE Series 246:

POWER AND IDENTITY:

TRIBALISM AND GLOBAL POLITICS

by

Harold R. Isaacs